Tsunami Man

Tsunami Man

Learning About Killer Waves with Walter Dudley

Anthony D. Fredericks

University of Hawai'i Press
Honolulu

Printed in the United States of America

07 06 05 04 03 02 6 5 4 3 2 1

Library of Congress Cataloging-in-Publication Data

Fredericks, Anthony D.

Tsunami man : learning about killer waves with Walter Dudley / Anthony D. Fredericks.

p. cm.

Includes bibliographical references and index.

ISBN 0–8248–2496–2 (paper : alk. paper)

1. Tsunamis. 2. Tsunamis—Hawaii. 3. Dudley, Walter C., 1945– I. Title.

GC221.2.F74 2002
363.34'9—dc21 2001052819

University of Hawai'i Press books are printed on acid-free paper and meet the guidelines for permanence and durability of the Council on Library Resources.

Designed by Santos Barbasa Jr.

Printed by Versa Press, Inc.

Dedication

To the memory of the teachers and students
who lost their lives at Laupāhoehoe School,
April 1, 1946

Contents

Acknowledgments

Tsunami Man: Learning About Killer Waves with Walter Dudley is a result of contributions from many people, beginning with Dr. Dudley himself. Throughout the interview process, he was gracious, cordial, and giving of both his time and research. He not only shared his life with me but enthusiastically contributed ideas, stories, and historical information, which added a personal touch to the scientific data. He is the utmost scientist, and I am proud to call him both friend and colleague. I am equally thankful for the contributions made by the staff of the Pacific Tsunami Museum—particularly Carrie Luke-Knotts and Noelani Puniwai, who provided personal insights into the work of Dr. Dudley, as well as important background information about Hawaiian legends and traditions. Their enthusiasm and knowledge contributed greatly to this book. I am equally indebted to the people of Hilo and Laupāhoehoe who shared stories about their lives and their tragedies. May they always be remembered.

Special thanks go to the reviewers of the initial manuscript who helped clarify significant points and ensure scientific accuracy. A special note of recognition to Dr. George Curtis for his talent and time in checking facts, and for his unselfish work in tsunami research; his contributions to the field are incalculable. Finally, I thank my editor, Keith Leber, for expertly guiding this book to completion. He is truly an "author's editor"!

Tsunami Man

The Monster Approaches 1

Splashes of early morning sunlight reflected off the bay's rippled water. The air was still and quiet; no breeze, no swaying of palms. An eerie silence spread across this picture postcard scene . . . a silence of anticipation, a silence of apprehension.

Walter Dudley waited on the roof of the Naniloa Hotel, overlooking Hilo (HEE-low) Bay, on the Big Island of Hawai'i. Eyes glued to the water, ears tuned to the ring of his cell phone, he was watching for a monster. It was a monster speeding across the Pacific Ocean at more than five hundred miles per hour. It was a monster with the potential for consuming lives and pulverizing buildings. It was a monster of enormous strength, incredible power, and unbelievable destructiveness.

Much earlier that morning, in an area northeast of the island of Hokkaido, Japan, an undersea earthquake had rocked the ocean floor. The sea floor had buckled and bulged, and a tsunami was generated. Oceanic waves spread out from the epicenter of the quake and began inundating coastal villages and small towns. As the ever-expanding waves raced across the ocean, countries were alerted and evacuation plans were put into motion. Using travel-time maps, scientists were able to predict the arrival of the waves at each island or nation. What they weren't able to predict was what the height of the waves would be when they hit the shore. The monster had been created . . . the monster was moving . . . but no one knew how big it was.

Walt Dudley waits on the roof of the Naniloa Hotel for the tsunami of October 4, 1994. Photo by John Coney.

As dawn unfolded on October 4, 1994, Walt Dudley waited and watched. He was well aware of the destructive tsunamis that had struck Hilo in 1946 and 1960, killing scores of humans and causing millions of dollars of property damage. Would this tsunami be equally destructive, equally deadly? Would it, once again, inundate Hilo's downtown area? Would it rumble across the tropical landscape and wash people and buildings out to sea?

Time seemed to stand still as Walt waited for the monster to arrive. He hoped it would be small. He hoped it would be quiet. He hoped the monster would simply go away. But for now, all he could do was wait as the unpredictable tsunami streaked across the ocean toward him and a city of thirty-seven thousand people. Walt Dudley waited and watched.

The monster approached.

2 The Making of Tsunamis

Tsunamis are born in the ocean. Some people call them tidal waves, but as you will discover in this book, they have little to do with the tides. *Tsunami* (pronounced soo-NAH-mee) is a Japanese word meaning "great wave in harbor." The word refers to a single wave or to a series of waves. Tsunamis are not dangerous at sea, but they have the potential for incredible damage when they go ashore. Other natural disasters, such as hurricanes and tornadoes, may be more common. But few can match the power or horror of tsunamis. They are truly "killer waves."

With 70 percent of the earth's surface covered by water, every continent's shoreline and all islands are at risk from a tsunami. Scientists estimate that there is an average of five to six tsunamis worldwide every year. Most are small, but a few are large and dangerous. With more and more people living in coastal communities, increased numbers of individuals are at risk from these potentially deadly natural phenomena.

To understand tsunamis, it helps to distinguish them from both tide- and wind-generated waves. Tidal waves are caused by the ocean tides, the rise and fall of water on a shoreline resulting from the gravitational pull of the sun and the moon. As the sun and the moon pull on the earth like magnets, the level of the earth's seawater is raised in some places and lowered in others. Twice each day, water comes farther up onto shore than normal, which is called high tide. At low tide, the

Great Wave off the Coast of Kanagawa by Hokusai.

water is pulled away from the shore and moves farther back into the sea.

The difference in height between high tide and low tide in a specific location is known as its tidal range. On open ocean coasts, the tidal range is typically between six and twelve feet. In some river mouths and bays, the tidal range can be much greater. For example, in the Bay of Fundy in Canada, the tidal range is an astonishing fifty-three feet—the greatest tidal range in the world.

Ocean waves, the second type of waves to be distinguished from tsunamis, are caused by the action of wind blowing across open expanses of the ocean. A series of small rounded waves known as ripples is set into motion as the wind blows across the water. And as it continues to blow, the ripples grow larger and turn into waves. These waves become longer and steeper as long as the wind blows. In addition, they continue to move across the surface of the water.

The highest point of a wave is its crest. Between the crests of two waves is a low valley called the trough. The height of a wave is the distance from the trough up to the crest. Waves usually follow one another, forming a "train" of waves. The time it takes two crests in a train to pass the same point is known as a wave period. Scientists use wave periods to figure out how fast waves are moving. Most ocean waves have periods of from two up to twenty seconds. This means that they never travel more than sixty miles per hour. The average speed of an ocean wave is about thirty-five miles per hour.

Near the shore, the sea bottom slopes upward. As an ocean wave approaches shore, the water becomes shallower. The wave slows down because the rising shoreline restricts the wave motion to a smaller and smaller volume, similar to when five lanes of heavy traffic are squeezed down to one lane of traffic. The cars must go slower to get through. Waves also begin to pile up as they approach the shallow shoreline. Eventually they topple over and crash onto the beach. This constant action is called surf.

A tsunami is different from a normal ocean wave, just as it is different from the rise and fall of the tides. A tsunami, unlike regular waves, is produced by one of three different types of violent geologic activity: submarine earthquakes, landslides, and volcanic eruptions.

Earthquakes: Most tsunamis are generated in the Pacific Ocean along the "Ring of Fire." This area is like a giant horseshoe extending from New Zealand in the southwestern Pacific, through Indonesia, past Japan and the Aleutian Islands in the north Pacific, and down the west coasts of North and South America. It is one of the most geologically active regions of the world. Many earthquakes and volcanic eruptions occur in this region every year.

Most tsunamis are generated when a section of the ocean floor is thrust upward or suddenly drops. About 86 percent of all tsunamis are the result of these undersea earthquakes. They usually occur along the edges of tectonic plates around the Pacific Rim.

The earth's surface is composed of approximately seven

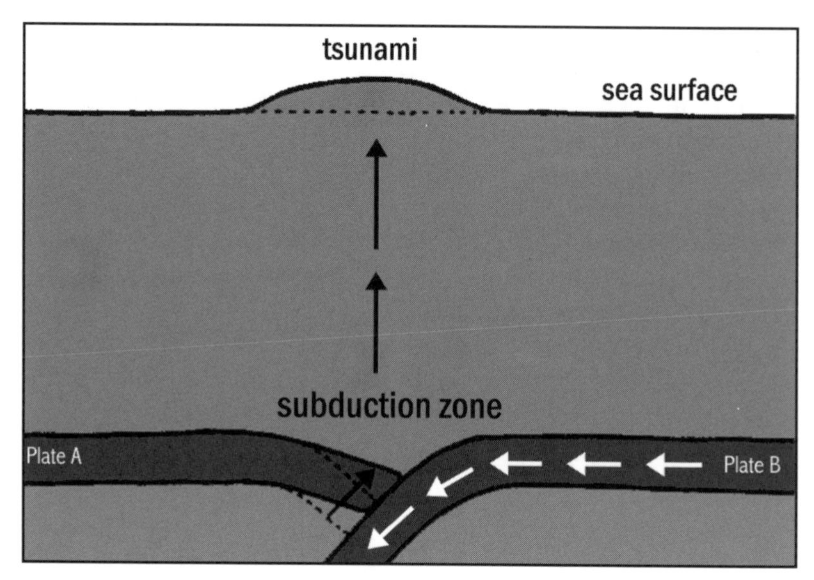

Illustration of an undersea earthquake. Plate *A* is slowly pulled into the subduction zone by plate *B*. Eventually, plate *A* snaps back (an earthquake) and generates a tsunami. Illustration by Amy Cutler.

large plates and several smaller plates. These tectonic plates—up to eighty miles thick—fit together like an enormous jigsaw puzzle. The plates are constantly moving, and sometimes the edges of the plates "catch" and pressure begins to build. When the pressure is too great, the tension is released as an earthquake. If the earthquake is powerful and occurs underwater, the ocean may suddenly shift and create a tsunami.

As the day drew to a close on July 17, 1998, on the northern coast of Papua New Guinea, families were preparing their evening meal. Children were playing on the beach. Neighbors and friends were chatting about the day's activities. The scene was peaceful and tranquil. But at 6:49 (local time), a magnitude 7.1 earthquake rocked nineteen miles of coastline. The ocean bottom was deformed, causing the sea surface to bulge upward. A tsunami was created. Three monstrous waves bore down on four coastal villages.

Striking within fifteen minutes of the main shock, the waves swept across the exposed beachfront. Residents had no time to escape. The waves, each up to thirty-three feet in height, carried houses and people into a nearby lagoon. Scores of people were impaled on broken tree branches. Others were buried under sand and debris. Many were crushed in their huts. Most drowned in the turbulent waters. Later, more than two hundred bodies were discovered one hundred miles to the west. It has now been estimated that more than twenty-two hundred villagers were killed by these monster waves. It was one of the deadliest tsunamis of modern times.

Landslides: The second most common cause of tsunamis is a landslide. Earth breaking away and falling from a mountaintop might plunge into the sea and generate a tsunami. Underwater landslides can also cause tsunamis.

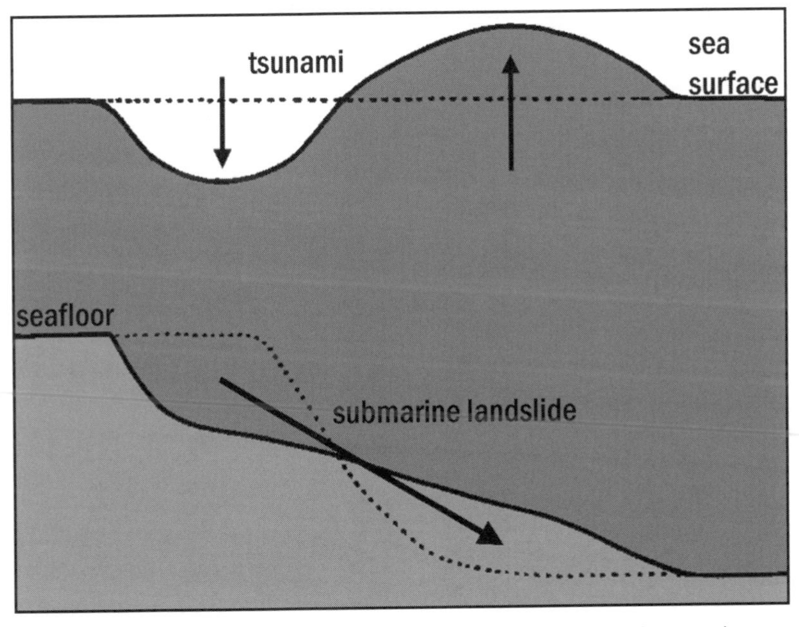

Illustration of a submarine landslide. A mass of material slides down a slope and generates a tsunami. Illustration by Amy Cutler.

Landslides occur when slopes become too weak to support their own weight. On land, this weakness is generally caused by rainfall or another source of water increasing moisture content of the soil. If a hillside is very dry, dirt and rocks can tumble down the grade. In certain places, large chunks of snow or ice may slide off mountains and crash into the sea. Underwater, earthquakes, storms, or an accumulation of sediment can cause slope failure.

In 1958, an earthquake measuring 7.5 on the Richter scale created a massive landslide near Lituya Bay, Alaska. A section of mountain fell into a nearby fjord, or narrow inlet, displacing an enormous volume of water. This water swept up the opposite mountain to a height of seventeen hundred feet, then rushed back down the mountain and into Lituya Bay. This sudden influx of water generated a terrifying tsunami. The wave rose to an estimated height of one hundred feet and sped across the tiny bay at a speed of a hundred miles per hour. Boats, buildings, and people were swept away in a matter of minutes.

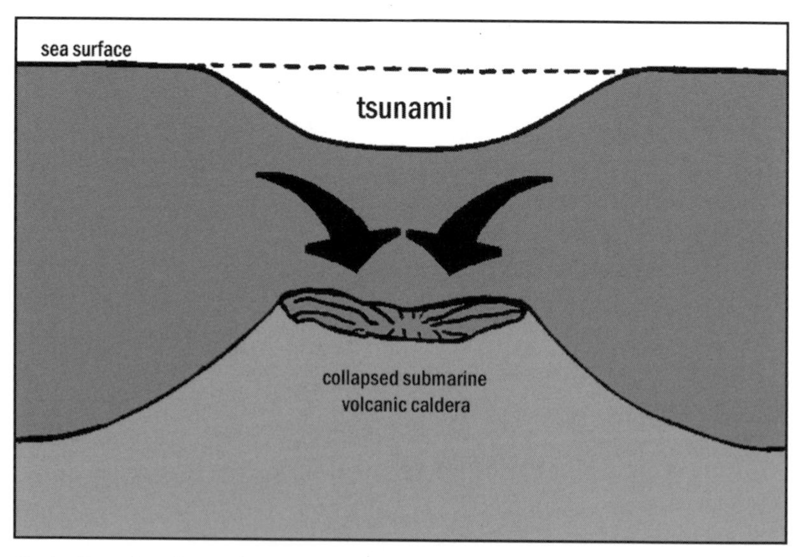

Illustration of collapsed underwater volcanic caldera. Volcanic material falls inward and generates a tsunami. Illustration by Amy Cutler.

Volcanic Eruptions: Near-shore or underwater volcanoes can also generate tsunamis. There are three types of tsunami-generating volcanic activities. The first is a submarine explosion—a violent release of magma underwater. The second is a sudden slip of material down the side of an erupting underwater volcano. The third type is when the caldera, or crater, of an underwater volcano is abruptly uplifted or depressed. In all three cases, the shape of the sea bottom is suddenly changed, causing the surface of the sea to change—often producing a series of waves.

On August 27, 1883, a violent volcanic eruption blew apart the tiny Indonesian island of Krakatau. The boom from the explosion was heard more than two thousand miles away. This powerful eruption triggered a tsunami with waves estimated at nearly 130 feet high. These walls of water surged along the coastlines of several countries. Hundreds of villages were wiped out. More than thirty-six thousand people were killed.

When a tsunami is generated, it begins to race across the open sea. Tsunamis are fast, very fast. They may travel through the open waters of the ocean at speeds of more than 500 miles per hour, about as fast as a jet plane! In 1960, an earthquake off the coast of Chile generated a tsunami. This tsunami traveled the sixty-six hundred miles between Chile and Hawai'i in just under fifteen hours. It sped across the ocean at an average speed of 442 miles per hour. Seven hours later (twenty-two hours after the earthquake) and ten thousand miles from Chile, twenty-five-foot waves washed up along the coast of Japan.

Tsunamis are not tall as they travel across the open ocean. They may be only a few feet from trough to crest. Most of the time, sailors at sea do not even notice them. In 1946, a freighter was anchored offshore from the city of Hilo, Hawai'i. Early on the morning of April first, the crew witnessed enormous waves crashing into the city and sweeping away everything in their path. Yet the men had felt nothing when the low waves had passed beneath their ship.

When a tsunami nears shore, the shallowness of the water acts like a brake. Suddenly the speed of a wave may drop from five hundred miles per hour to one hundred miles per hour. The water beneath the wave piles up. Crests rise higher and higher. In seconds, a two-foot-high wave at sea may become thirty feet high on the shore. Typically, there are several waves in a row—a tsunami wave train—and the later waves in the train jam together and build up even higher than the first. It's not unusual for the third or fourth wave in a train to be the highest, but it may sometimes be the eighth or ninth wave. A tsunami can last for eight hours or more. Unlike some natural disasters, a tsunami might be dangerous for a long time.

Also unlike hurricanes or tornadoes, tsunamis aren't seasonal. They can occur at any time of the year, in any ocean. They can occur on bright sunshiny days or in the middle of the night. Scientists such as Walt Dudley are learning more and more about tsunamis all the time. What they learn is helping to save lives around the world.

3 Walter Dudley: Tsunami Man

One of the world's most destructive natural forces, tsunamis are the stuff of legends, a drama of nature, and a phenomenon both ancient and new. But for Dr. Walter Dudley of the University of Hawai'i at Hilo, tsunamis are his life's work. It is this work that has expanded our knowledge of these waves and helped people throughout the world better prepare for these unpredictable, yet ever-present, dangers.

Walt Dudley lives near Hilo, Hawai'i, with his wife and four children. He has been interested in the ocean since he was a child.

"When I was growing up in North Carolina," Walt says, "I would spend my summers down at the coast where my grandmother lived. She would tell me stuff about the ocean, and I really gained a love and fascination for the sea."

Walter C. Dudley was born in 1945, in Greensboro, North Carolina, about 230 miles from the Atlantic Ocean. When he was six years old, his family moved to Charlotte, North Carolina, which was even farther away from the ocean. Summers with his grandmother were his favorite time of all.

"My grandmother knew a lot of lore, a lot of natural history that she learned on her own. She'd also been to some summer marine schools in Maine. My grandfather had been a fisherman off the coast of North Carolina. So, I suppose, that's where my real love of the ocean came from."

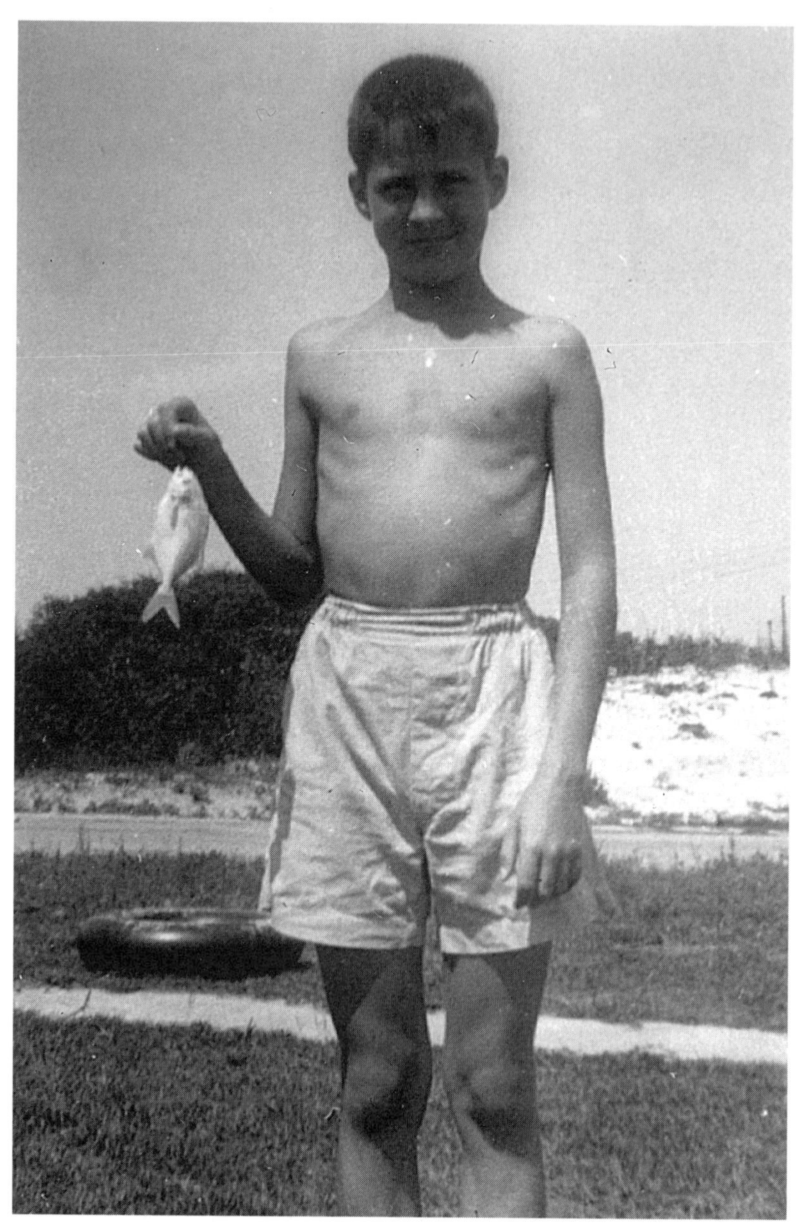

Walt as a young boy, in North Carolina. Photo courtesy of Walter Dudley.

When Walt graduated from high school in 1964, he went to the Northwestern University near Chicago as a National Merit Scholar. At Northwestern, Walt took a variety of courses, including engineering, theater, music, anthropology, art, and geology. Eventually, Walt obtained a degree in geology, because "that was as close to studying the ocean as I was going to get at Northwestern." During his summers at home, he worked as a swimming instructor and lifeguard.

After he graduated from college, he decided he wanted to live in a warm climate and attend graduate school. He chose the University of Hawai'i because of the weather. It was also one of the leading oceanographic institutions in the country.

Over the course of the next several years, Walt enlisted in the army, returned to Honolulu to finish his doctoral degree at the University of Hawai'i, and was awarded a National Science Foundation fellowship to France. After a year of research in a scientific laboratory, he was offered a job with the French Atomic Energy Commission. "I was studying plankton to try to find out the climatic history of the Ice Age, and I went on a fascinating research cruise with the French in the Atlantic Ocean, mapping parts of the sea floor. We dredged up the insides of undersea faults, and all the scientists were going nuts because we were bringing rocks up onto the deck that they had never seen before. That was very exciting."

Besides the scientific work he does today, Walt also enjoys a number of other activities. Having taught himself to play the guitar, he occasionally sings in local clubs. An avid surfer, he frequently goes out in the early morning to ride the waves before heading off to the university. He learned to ski when he was young, and he now takes a skiing trip to Utah every spring with his children. His one true passion, however, is karate. He has earned his black belt and occasionally teaches classes at the karate club in Hilo.

"I really like karate because it's a good all-around workout . . . it's serious stuff. I've broken some bones and not even realized they were broken," Walt says.

Walt and his daughter Malika at the karate studio in Hilo. Photo courtesy of Walter Dudley.

When Walt was in graduate school, he had the opportunity to teach some courses for his professors. He enjoyed the people and sharing his love of and excitement about science with them. Walt comes from a family of educators; his mother was a college librarian, and several other relatives were teachers, principals, or school superintendents. Although Walt loved the experiments and the research involved in his scientific work, he enjoyed the contact with people most.

In 1980, a teaching position opened up at the University of Hawai'i campus in Hilo. Walt jumped at the chance to begin a marine science program. As the only oceanographer on the university teaching staff, he was in charge of courses on the chemical, geological, and biological components of the world's oceans. Although he'd taught a lot of science, Walt discovered that what really captured his students' attention was when he talked about things that happened to people.

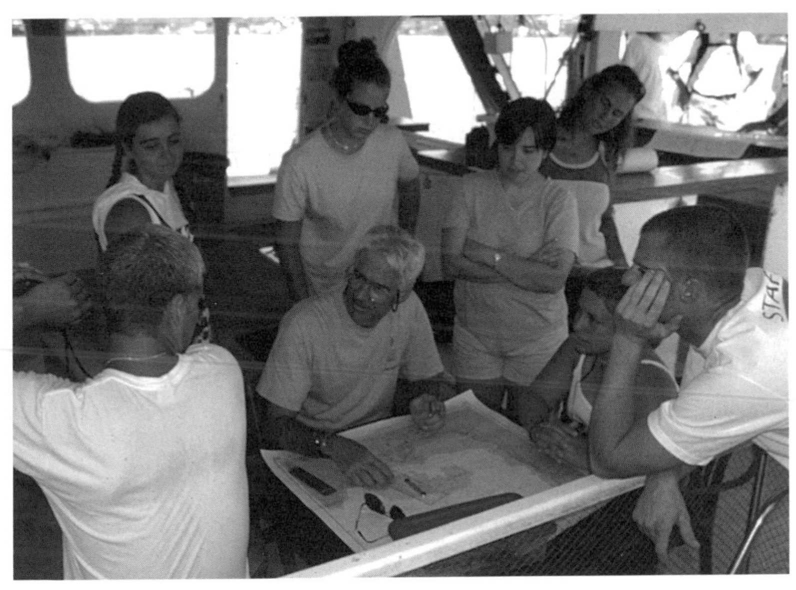

Walt works with students on a boat near the University of Hawai'i, Hilo. Photo by John Coney.

Walt says, "It's the stories about real people that make science come alive. Stories get everyone's attention. Students are motivated to learn science because it is personal; it's real. With stories, students say, 'Wow, science is really neat!'"

In gathering information for his courses, Walt learned about the disastrous tsunamis that had struck Hilo in 1946 and 1960 and how they had affected so many people and changed so many lives. He discovered that Hilo had recorded more tsunamis than any other place in the world. The information was particularly valuable because it was part of the personal history of hundreds of Hilo citizens.

As both a teacher and a scientist, Walt wanted to share the facts about tsunamis. But he discovered that when he wrapped the scientific data around stories of past tsunamis and their effect on real people, it became more meaningful. So Walt began talking to tsunami survivors. He believed that their stories would add an important human dimension to the scientific data, and that their stories might also save lives.

4 Gathering the Memories

While talking to the elderly survivors of the 1946 tsunami, Walt realized that when these people died, their stories could be lost.

"I realized that I was talking to people about history and that nobody was saving any of these stories. This local history has the power to teach people what they need to know: how to prepare for another tsunami. At some point I realized I was the only guy doing this. I began to feel this tremendous responsibility."

Walt understood that the stories people told were often about the single most important event in their lives. Many people lost their homes and businesses. Several lost their friends and family members. Some stories were not easy to tell simply because of the death and destruction that accompanied them.

"Ronald Yamaoka was twelve or thirteen when the tsunami of 1946 struck Laupāhoehoe School," Walt told me. "He remembers the tsunami washing over the school grounds and knocking people down and washing them out to sea. He was washed out to sea, too, and floated among the debris. After a long time and just by luck, a boat came by and picked him up. One of his school friends was lost, killed by the powerful waves. Mr. Yamaoka talked with tears in his eyes and sadness in his voice."

One of the things Walt is quick to point out is that he's not a "lab-coat scientist." By that he means he doesn't spend a lot of time in a laboratory analyzing seismographic data or surveying inundation zones. Scientists such as George Curtis at the University of Hawai'i's main campus in Mānoa and Frank Gonzales at the National Oceanic and Atmospheric Administration (NOAA) Pacific Marine Environmental Laboratory in Seattle, Washington, have conducted extensive research on tsunamis and how they behave. Walt believed he could combine this information with the stories of tsunami survivors and do something nobody had ever done. The scientific facts were important, but it would be the stories of tsunami survivors that would teach people what they really needed to know.

Walt spends a lot of time recording survivor stories. People write to him and tell him they survived a tsunami. Sometimes children of tsunami survivors contact Walt by phone or by mail to tell him that a parent has a story to share. Walt might read in the newspaper about an individual who survived a tsunami, or someone who knows a survivor. Often survivors call Walt directly.

Walt works with his team to set up the necessary equipment on a boat in preparation for an interview with a tsunami survivor. Photo by John Coney.

When he learns about a person with a story to share, Walt contacts them by letter or phone to arrange an interview. If the person is on the Big Island of Hawai'i, Walt will drive to that person's house to speak with him or her. If the individual lives on another of the Hawaiian Islands, Walt will fly to that island with his tape recorder, notebook, video camera, and still camera.

Walt also enlists Jeanne Johnson to help with the interviews. Jeanne is a survivor of the 1946 tsunami and is nearer the age of other tsunami survivors than Walt. Jeanne sometimes makes the initial contact, sets up the appointments, and asks some basic questions during the video interviews. Walt often sits behind her, analyzing the interview and occasionally asking questions to fill in gaps or keep the individual on track. During every interview, Jeanne and Walt encourage each person to tell his or her own version of the story, just as it was experienced.

One woman told how she miraculously survived the tsunami of 1960. She was standing in her house shortly after midnight on May 23. Suddenly, she heard a loud explosion and was enveloped in darkness. Before she could think, a wall of water rushed into the house. She and all her possessions were spun around in a whirlpool of incredible power. She was hit in the head and lost consciousness. The next thing she knew, she awoke tangled in a mass of bushes. Seconds later, the wave sucked her back into the water, where she floated on her back among a mass of debris.

The night was pitch black—no lights anywhere. Suddenly, and without warning, she was swept up by another large wave. She swallowed large amounts of water and gasoline. She grabbed something floating by. She then found herself being pulled out into ocean, far from the city of Hilo. The ocean's surface was littered with debris—parts of houses, scraps of wood, and tons of garbage. She realized that what she was clinging to was a screen door from her own house.

All night long she held fast to that door. She was tossed by the turbulent waters and feared she might drown or be attacked by sharks. She was at the mercy of the tides and currents flowing around the Big Island of Hawai'i. As morning dawned, the tide slowly pushed her back toward the city of Hilo. After what seemed like the longest night of her life, she finally spotted a rescue boat approaching. Two sailors jumped into the water and helped her back to the boat. Her only injuries were a cut finger and a bruised knee. But for a long time afterward, she could not bear to hear the sound of running water. She had survived one of Hawai'i's greatest natural disasters, but it would forever be a terror in her life!

Walt and Jeanne always get permission from each person they interview to use the story. They also ask to tape record or videotape the interview. These recordings are important in helping to accurately preserve the recollections.

After each interview, Walt turns the audiotape over to a transcriber, a person who listens to the tape and types up the story exactly as it was told. The transcription is then filed with the stories of other tsunami survivors. Eventually, it might become

Preparing for an interview with a tsunami survivor on the island of Moloka'i. Photo by Walter Dudley.

part of an exhibit at the Pacific Tsunami Museum. It could be included in a section of a book or other publication Walt writes. Most important, though, the story has been preserved for another generation.

Sometimes organizations such as the National Geographic Society want to interview tsunami survivors for a video or news story. Because many tsunami survivors are getting old, their memories may not be as complete as they once were. Others would rather not share their stories over and over again and would like to maintain their privacy. With the permission of these interviewees, Walt is able to provide the organizations with copies of his interviews. This is Walt's way of protecting the people who have shared an important and very personal event in their lives.

Walt knows that the best information about tsunamis comes from the people who have lived through these terrible disasters. It is their stories that put a human face to the scientific facts and figures. So Walt continues to talk to the survivors of tsunamis. He knows that the real power of a tsunami is in what the survivors' stories tell people to help better understand how to prepare for another one in the future.

"Talking Story"

It is believed that about A.D. 300, people from the Marquesas Islands discovered an isolated group of islands in the North Pacific Ocean that had existed for more than seventy million years. These ancient voyagers sailed across twenty-four hundred miles of uncharted ocean to make the islands their home.

Around A.D. 1200, a second group of Polynesian seafarers—this time from the Tahitian Islands—sailed nearly twenty-seven hundred miles in search of a new home. Using the stars to guide them, these adventurers made several voyages across the ocean. The Tahitians settled in the islands and gave them the name of their old home, Havaiki. That name was eventually transformed into what we now call these islands, Hawai'i.

These ancient peoples brought many things to their new land. Seeds, animals, and tools traveled across the vast expanse of the ocean. So, too, did the legends that were part of the language and lives of these brave explorers. These tales were of ancient times, or long voyages, and they were a way of preserving and celebrating the peoples' heritage.

The stories were also a record—an oral record—of the history of the ancient Hawaiians. Each story was passed from generation to generation through chants and through hula, the traditional dance that tells a story.

When the missionaries arrived in Hawai'i, in the nineteenth century, they taught the native Hawaiians how to read and write. Until that point, there had been no written language. As part of their teachings, the missionaries prohibited the telling of ancient stories. But Hawaiians practiced their storytelling in private and in secret ceremonies. It was a part of their heritage they were unwilling to give up.

Today, this practice of storytelling—now known as "talking story"—is one way native Hawaiians still preserve and share their history. More than just factual history, it is often personal recollections of the members of a specific family; the funny anecdotes, humorous stories, and fond reminiscences about relatives that are shared during family gatherings. Because Hawai'i is such a cultural melting pot, "talking story" is also a way of preserving qualities from each culture while blending them into a composite story. "Talking story" is similar to the conversations friends might have at a party.

"Talking story" is a distinctively Hawaiian tradition that celebrates and enriches the lives of families, while preserving both personal and cultural history and bonding one generation to the next.

5

Tsunamis in Hawai'i

At least a dozen tsunamis have struck the Hawaiian Islands since the early nineteenth century, when missionaries began keeping written records. Interestingly, more people have been killed by tsunamis in Hawai'i than by hurricanes, earthquakes, and volcanic eruptions combined. Here are some memorable ones.

November 7, 1837

A brilliant sunrise etched its way across the bay at Hilo. Nearly ten thousand people were assembled on the beach for a daylong religious ceremony. There were laughter and song in the air; it was a grand and glorious celebration. Suddenly, the ocean receded. A large part of the bay was completely exposed. Many people, not knowing what was happening, rushed down to witness this strange sight. Before they realized it, an enormous wave had formed offshore and was speeding directly at them. According to one witness, "The sea . . . had all of a sudden risen in a gigantic wave, and this wave, rushing in with the rapidity of a racehorse, had fallen upon the shore, sweeping everything into indiscriminate ruin. The sea crashed upon the shore as if a heavy mountain had fallen on the beach."

The water surged and roiled up the beach. The powerful wave carried away dozens of homes. Men, women, and children were swept into the bay. People were tossed about as though they were toys. Many families were separated, and buildings collapsed.

Even the strongest swimmers were sucked under the water from the powerful force of the waves. In the end, more than a dozen people lost their lives. Devastation and destruction lay all around.

May 10, 1877

On the evening of May 9, the earth shifted off the coast of Chile. A sudden and enormous earthquake shook the undersea floor. A large mass of water was displaced, generating a tsunami. The waves raced across thousands of miles of the Pacific Ocean. Eventually, they came ashore at Hilo, in the early morning hours of May 10.

Most people were asleep when the waves careened through the middle of the city. Large buildings were washed away. Giant boulders were deposited on the streets. Homes were tossed into the ocean as if they were weightless. Stores, bridges, and vessels

Coconut Island, a small island that juts out into Hilo Bay. A small hospital used to be on this island. Photo by author.

were swept away as though by a giant hand. A large sailing ship that had been anchored in twenty-four feet of water in Hilo Bay was left high and dry as the water receded out to sea. When the ocean rushed back in again, the ship was spun around and around like a toy boat.

Coconut Island, which juts out into the bay, was swept clean. A small hospital on the island was torn from its foundation and disappeared beneath the water. Houses were lifted and scattered across the countryside. Buildings were ripped apart and washed away. Throughout the town, there was a general state of panic.

The first rays of daylight revealed the destruction everywhere. Buildings were gone, dead or dying livestock lay scattered about, and scores of homeless people staggered through the ruins. The waves had risen to a height of more than sixteen feet above sea level. Although only five people lost their lives, the destruction would be remembered for many years.

April 1, 1946

In Hawai'i, it was almost two hours past midnight when somewhere off the southern coast of Alaska, the earth shuddered. A rolling underwater earthquake along the Aleutian Trench ruptured the sea floor and displaced a huge amount of seawater. Hawai'i was more than two thousand miles away, but in less than five hours it would experience one of its worst natural disasters.

That Monday morning dawned bright and clear for the students at Laupāhoehoe (lah-PAH-hoy-hoy) School, on the northeast edge of the Big Island of Hawai'i. The school sat on a small peninsula. Many students were gathered along the seawall fronting the baseball field. Others were playing and laughing. Some were still arriving by bus or walking along the road that snaked down from the cliffs above. All were filled with anticipation of the lessons and activities that awaited them in the nearby classrooms.

At ten minutes to seven, the scene changed dramatically. The sea receded, far beyond the jagged rocks that encircled the tiny peninsula. Awestruck students ran down to get a better view.

Laupāhoehoe School on Laupāhoehoe Point before the tsunami of 1946. The four teachers' cottages are nearest the ocean. The athletic field is immediately behind the cottages. The school buildings are behind the athletic field. Photo courtesy of the Bunji Fujimoto Collection, Pacific Tsunami Museum.

They couldn't believe what they were seeing. Without warning, the sea began to swell. Water surged above the rocky shoreline, rising higher than it ever had before.

The ocean receded and then rose in another wall of water. Once again, the water retreated. Nearly a half hour after the first wave, the third wave arrived. This wave was larger and more powerful than the first two. Children were knocked off their feet and swept out to sea. Water came in from all sides of the tiny peninsula. Water rushed through the teachers' cottages, over the athletic fields, and into the agricultural building. The wave collapsed the grandstand, filled with students. Elsewhere, children ran away from the sea, trying to distance themselves from the turbulent waters. Some were tossed into bushes as the sea seethed all around them. Others were sucked back into the ocean with such power, they were never seen again. Cries and screams filled the air.

These tsunami waves were devastating. The third rose more than thirty feet above sea level, leaving a path of destruction. Where once stood a vibrant and active school, there lay a tumble of buildings and a scattered wasteland of debris. In the end, twenty-four people were killed, including sixteen students and four teachers.

The marble monument that now stands at Laupāhoehoe Point commemorating the students and teachers who lost their lives in the tsunami of April 1, 1946. Photo by author.

Today, a solemn marble marker stands at Laupāhoehoe Point. This stone is a silent reminder of the power of the sea. More important, it memorializes the young lives taken by this awesome force.

May 23, 1960

On Sunday, May 22, the South American country of Chile was being rocked. It was being shaken and rattled by one of the largest earthquakes in recent memory. Tremors rolled through the ground as the earthquake intensified, culminating in a monstrous shaking that lasted for more than fifteen minutes.

The earthquake registered a staggering 9.6 on the Richter scale. In turn, a tsunami was generated, a series of waves that sped across the expanse of the Pacific Ocean. The tsunami headed directly toward the city of Hilo, Hawai'i, where it would take less than fifteen hours to arrive.

Not knowing how big the tsunami would be, the Pacific Tsunami Warning Center near Honolulu issued a warning at 6:30 P.M. Sirens throughout the Hilo area began to sound at 8:30 P.M. Some people started packing their things and loading up their cars. Business people gathered important documents and prepared for evacuation. Police were mobilized and evacuation centers were opened. But few moved out of town. Many did not understand the meaning of the sirens. Radio broadcasts carried conflicting information. Several people did not believe that this alarm was a serious one.

Everything changed shortly after midnight. Water in the bay rose to a height four feet above normal, then fell to three feet below normal. The first wave swept through Hilo Bay. Fifteen minutes later, the second wave, more than nine feet high, spilled over the seawall and through the business district. Soon afterward, the water withdrew once again. This time it fell to a level more than seven feet below normal. The worst was yet to come.

Shortly after 1:00 A.M., a thirty-five-foot-high wall of water roared into town. The power plant was flooded, sending an explosion of sparks into the night sky. Bridges, homes, and businesses

Parking meters in Hilo are bent over by the force of the waves of the 1960 tsunami. Photo from the U.S. Army Corp of Engineers, provided by the National Geophysical Data Center.

were carried away. Water raced up the streets, flattening buildings and tossing cars around like matchsticks. Twenty-two-ton boulders were wrenched from the bayfront seawall and rolled more than six hundred feet inland. Parking meters were bent over parallel to the ground. An eleven-ton tractor was swept out into the street. Steel buildings were twisted and torn from their foundations. Entire blocks were flattened, leaving gigantic piles of wreckage throughout the town.

In the aftermath, Hilo looked like a war zone. More than five hundred buildings, from small one-bedroom houses to enormous steel and concrete structures, were demolished. Sewage and other rubbish were scattered everywhere. More than 580 acres of land

The Waiakea town clock, frozen at 1:05 A.M. when the largest wave struck Hilo on May 23, 1960. This clock is a monument to the 1960 tsunami. Photo by author.

were inundated by the tsunami. Even longtime residents had difficulty recognizing sections of town—they had simply disappeared. Property damage was estimated at more than fifty million dollars. Sixty-one people died and nearly five hundred people were injured. The power of the tsunami was unbelievable. It was definitely unforgettable.

Tsunamis are a fact of life in Hawai'i. When will the next one arrive? No one, not even Walt Dudley, knows the answer to that question.

An Ancient Legend

The ancient people of Hawai'i shared many legends about their islands. Some legends were based on observations; others were created to explain the mysteries of nature. Since the ocean has always been an important part of Hawaiian life, many stories are told of its power and influence.

One legend from long ago tells of a time when the people of the island of Hawai'i forgot to thank their gods for the food the people grew and the fish they caught. One man, a poor fisherman, grew angry when he could catch nothing. Shouting at the gods, he said, "You do not care if the poor fisherman dies. I say you are not gods! If there are indeed gods in the sea, then let me see you!"

At once the sky darkened and the waves began to surge back and forth. One of the ocean gods, the shark chief, rose from the waters and said that the people of Hawai'i had forgotten who provided for them. To teach them a lesson, the ocean would roll over the island of Hawai'i and cover it. The fisherman asked the shark chief for his forgiveness, saying that he would remember the gods with prayers and offerings. The shark chief took pity on the poor fisherman and told him to take his wife and climb to the highest mountain peak.

The fisherman paddled quickly to shore and ran to tell his wife of the shark's words. They hurried up the side of Mauna Kea (MAH-nah KAY-ah), scrambling over the rough rocks and sharp grasses. When at last they reached the top of the mountain, the woman began a slow chant:

"O Hawai'i, my home,

I shall see you no more!

The many-colored sea will cover you—

The blue sea will cover Hawai'i."

The man and his wife heard a distant roar and saw a wall of water rushing toward the island. The water cascaded over the beaches and into the valleys, carrying houses and trees before it. Wave after wave rose up and covered the landscape with its terrible power. Only the man and his wife were left, on a small island of rock in the middle of the waves. After a long sleep, they awoke to find that the ocean had returned to its rightful place, but the entire island was barren, swept clean of homes, vegetation, and people. It would be many years before Hawai'i would again be a land of green forests and lush gardens. It would also be a land where the people never again forgot their gods.

6

A Tsunami Day

At 3:23 A.M. (Hawai'i time) on October 4, 1994, the earth hiccups. An earthquake measuring 8.2 on the Richter scale rumbles near the Kuril Islands, a group of small islands northeast of Japan. The sea floor above the earthquake's epicenter rises by as much as six feet, causing a large bulge of water to form.

A monster has been created! The monster races across the northern Pacific Ocean, inundating coastal towns and washing houses and buildings out to sea. It spreads out across the ocean at breakneck speed. Hawai'i is directly in its path.

4:23 A.M.: The Pacific Tsunami Warning Center in Honolulu issues a Pacific-wide tsunami warning. Alerted to the earthquake, scientists notify officials throughout Hawai'i, as well as those in California, Oregon, Washington, Alaska, and the west coast of Canada, that a tsunami is imminent. Emergency personnel are called and evacuation plans are put into action.

6:30 A.M.: The Hawai'i State school superintendent is contacted. A decision is made to close all 238 public schools as well as the 9 campuses of the University of Hawai'i. Civil defense officials decide to issue the alarm an hour earlier than required. This will help prevent major traffic problems. Sirens along the coast are sounded and evacuation shelters are opened. Residents are advised to proceed to higher ground. Tourists and vacationers are encouraged to move to the upper floors of hotels. Walt Dudley is awakened by a phone call. He climbs out of bed and turns on the radio.

7:19 A.M.: Registering a mere six and a half inches on the tide gauge, the tsunami passes by Wake Island. Scientists know that a half-foot increase at a distant island could turn out to be a major increase in the Hawaiian Islands. They don't want to take any chances. Preparations are speeded up: roadblocks are set up, portable telephones are tested, volunteers ready themselves, and buildings are vacated. People are moving quickly, but there is no panic.

Walt loads his car with camera equipment and scientific instruments. He drives to Banyan Drive, where the resort hotels are located. There is little traffic on the streets as he arrives at the

The Naniloa Hotel on Hilo Bay. Walt Dudley sets up his monitoring and photographic equipment on the roof of this hotel in advance of an approaching tsunami. Photo by Malika Dudley and Emily Dudley.

Naniloa Hotel. After showing his civil defense badge, he is waved through the barricade and into the inundation zone. He parks his car and begins unloading it. He's not sure whether it will be swept away by the force of the approaching tsunami. But he's calm and knows what to do as the monster approaches!

8:00 A.M.: The tsunami reaches Midway Island, the first inhabited island in the Emperor Seamount–Hawaiian Island Chain. This string of islands is about thirty-five hundred miles long. The tsunami is racing across the ocean at nearly five hundred miles per hour and will strike Hilo in a matter of hours. At Midway, the tsunami registers twenty inches on the tide gauge. This could be the precursor of a major tsunami in Hawai'i. No one is sure.

8:02 A.M.: Walt finishes unloading the equipment from his car. He lugs his radios and cameras up to the roof of the hotel. He passes by dozens of hotel personnel carrying computers, hotel records, and other items outside. A fleet of vans is loaded with articles of every size, shape, and description. Each is ready to speed along Banyan Drive and onto Kanoelehua (kah-no-eh-lay-WHO-ah) Avenue to higher ground. People are moving quickly.

9:00 A.M.: Walt works with his colleague John Coney to set up his video and time-lapse movie cameras on the roof of the Naniloa Hotel. He establishes radio contact with another colleague, George Curtis, who is on the roof of the Bayshores Towers Condominium, across the bay. Walt receives a call from Bruce Turner of the Pacific Tsunami Warning Center (PTWC). Bruce shares his concern about the size of the wave at Midway Island. He thinks that the tsunami will be large and powerful. Using the PTWC computer, Bruce predicts that the first wave of the tsunami will strike Honolulu at 10:42 A.M. They estimate that it will come ashore at Hilo at 11:08 A.M. The tsunami is a little more than two hours away, but it's still traveling at nearly five hundred miles per hour. Its speed doesn't tell them how big it will be when it arrives. All they can do is wait.

The Pacific Tsunami Warning Center, located on the island of O'ahu. Photo by Walter Dudley.

9:30 A.M.: Throughout Hawai'i, preparations are at a fevered pitch. Roadblocks are set up to prevent people from heading to the shore. Hotels are moving tourists from lower floors to higher floors. Special buses are carrying people from low-lying areas to special sites away from the beaches. Many beaches are closed, and numerous tourist attractions are shut down. People are filling pillowcases with sand to create emergency sandbag barricades. Others are fortifying their homes with sheets of plywood nailed across doors and windows. Business people are hurriedly loading their cars with important business records and computer files. Dozens of boats, from tiny sailboats to large cargo ships, head out of every bay and harbor to anchor far offshore. In all seaside cities in Hawai'i, there are beehives of activity.

9:45 A.M.: Walt receives a call on his cell phone. A local radio station wants to interview him about the approaching tsunami. As a news media contact for the International Tsunami Information Center, it is Walt's job to provide data to television and radio stations. This helps the public get the accurate information they need, ahead of time. During the interview, Walt stresses the seriousness of the situation and why people must act quickly. He hopes that nothing will come of the warnings, but he has to make sure everyone is prepared. After the interview, Walt checks his equipment once again. He is ready.

10:15 A.M.: There is a strange silence in the air. Walt waits and watches. His cell phone and radio are quiet. The only sounds he hears are his own breathing and the distant screech of seabirds. There is an eerie calm in the air. All Walt can do is wait and watch, wait and watch. The monster is coming.

10:20 A.M.: The tsunami reaches the first major Hawaiian Island, Kaua'i (kah-WHY-ee). There is a run-up of fourteen inches.

10:30 A.M.: The tsunami passes by the island of O'ahu (oh-AH-who) with a run-up of twenty-two inches. Walt is getting concerned; it seems each island in the chain is experiencing a larger run-up. O'ahu is the third island. The Big Island of Hawai'i, where Hilo is located, is the eighth major island in the chain. How large will the tsunami be when it reaches Walt's position? What will it do? How much damage will it cause?

10:44 A.M.: The tsunami reaches the island of Maui (MOW-ee). It has a reported height of nearly three feet. Each reported wave is larger than the preceding one. Walt checks his cameras again. He looks out over Hilo Bay. Everything is quiet; everything is still. "What's going to happen?" he asks himself.

10:50 A.M.: Carefully scanning Hilo Bay, Walt notices the water take on a peculiar, dull look. Glancing up the Wailoa (why-LOW-ah) River, he sees a muddy plume surging into the bay. A series of

A view of Hilo Bay. Photo by author.

large ripples begins moving back and forth across the water. Walt watches very carefully. Does this mean a devastating wave is about to strike? Is the water in the bay about to be sucked out? Is this the monster? Scientific instruments in Hilo Bay record a wave height of less than one and a half feet.

11:08 A.M.: The predicted arrival time for the tsunami comes. There is little change in the water.

11:10 A.M.: The predicted arrival time passes. No large wave appears. Would there be a second wave? Would a third and even larger wave follow? Walt has many questions but no clear answers as he continues to scan the bay. Then he watches as a solitary sailboat, which had been drifting with other boats outside the breakwater, heads back into the bay. "What is it doing?" Walt

A sailboat ignores warnings from a Hawai`i county helicopter during the tsunami warning of October 4, 1994. Photo by Walter Dudley.

wonders. "The skipper is crazy." The warning is still in effect; the danger has not passed. Walt watches incredulously as a Hawai'i county helicopter circles over the yacht to warn the skipper that the danger is not over. But the skipper simply ignores the warning and continues his journey back into the bay. Walt thinks, "If a significant tsunami wave strikes at this moment, that sailboat will be dashed against the rocks along the shore. There will be no hope of survival."

11:55 A.M.: The warning is canceled. No destructive waves have been reported in Hawai'i. Walt and thousands of other people throughout the Hawaiian Islands breathe a collective sigh of relief. As Walt begins to pack up his gear, he wonders when he will be climbing onto the roof of the hotel again. When will he be scanning once more the seemingly placid waters of Hilo Bay? When would an oceanic monster come and strike his tropical paradise? What would it look like? What would it do? These are questions he continues to ask. He doesn't know the answers, but he is ready. He hopes everyone else is, too.

Separating Fiction from Fact

Many people misunderstand tsunamis. Walt's job is to provide to the public correct information about these natural phenomena. The information helps us understand their power and is helpful in preventing the loss of lives and property when a tsunami does strike. This is why Walt spends a lot of time on Hawai'i television and radio stations talking about these killer waves.

Most people think a tsunami is one giant wave that towers ominously over the shoreline. This kind of wave does occur, but it is a rarity rather than a regular formation. As Walt points out, "A tsunami typically consists of ten or more waves forming a 'tsunami wave train.' The individual waves follow one behind the other, anywhere from five to ninety minutes apart." It is not unusual for a tsunami wave train to last eight hours or more.

Illustrations of shorelines where the water has receded in advance of a tsunami have given the impression that a large drop in sea level precedes every tsunami. This is not always true. When Walt talks to his students at the University of Hawai'i, he explains that a rise in water level may precede a tsunami, just as a decrease in water level may. It depends on what part of a tsunami wave train reaches shore first. If a wave trough does, there will be a sudden decrease in water level. But if a wave crest comes ashore first, there will be a rise in the water level. Walt is quick to point out that any sudden increase or decrease in the level of the ocean should be interpreted as a danger sign, a signal that a tsunami may be imminent.

Some people think that because tsunamis can be large waves, they are ideal for surfing. Nothing could be further from the truth. Walt notes that tsunami waves are not surfable waves. Most tsunamis are more like floods—rapidly moving water filled with debris, sometimes as large as houses. The water churns furiously and is extremely dangerous. Trying to surf a tsunami is comparable to a mouse trying to swim in a washing machine. It would be a deadly experience.

Inundation of coastal property on the island of O`ahu during the tsunami of 1957. Photo by Henry Helbush, provided by the National Geophysical Data Center.

7
Sharing the Wisdom

Tornadoes, hurricanes, forest fires, and other natural disasters can often be predicted. People are alerted and can be removed from harm's way. In Florida, people have learned how to move from the path of a hurricane. In Western states, people are evacuated from areas threatened by forest fire. There are often lots of clues to an approaching storm.

"In the case of tsunamis, it may be a beautiful day at the beach, the sun may be shining, and the temperature warm. In other words, there are no external signs that anything is wrong," says Walt. "But people have to understand that all this can change very quickly. Many people still believe that tsunamis are related to the weather. They believe that tsunamis can be predicted just as easily as an advancing line of thunderclouds. They think that if it's a warm and cloudless day, then there's no danger of a tsunami. That's wrong and it's also dangerous."

Informing the public about the dangers of tsunamis is sometimes challenging work. Walt must share complex scientific information with the public in a way that will prevent loss of life.

Walt knows that most people, including most people who live in Hawai'i, have never seen a tsunami. Their knowledge about tsunamis often comes from movies or novels. Unfortunately, much of that information is wrong, because it has been sensationalized to add adventure and extra thrills to a story. Some people even believe a tsunami is itself fictional. Because tsunamis are infrequent, many people think they are less dangerous than the more common hurricane or earthquake.

"The West Coast of the continental United States has a serious threat from tsunamis, and the East Coast has a potential one. In Hawai'i there is an ever-present threat. So, one of my jobs is to let people know that tsunamis are nothing they have to be afraid of, but something they have to learn about. As long as people know what to do, they are going to be okay. The more people know and understand tsunamis, the safer they will be," Walt says.

During the past few years, Walt has spent a lot of time working with the news media, talking with newspaper reporters, radio broadcasters, and television announcers. He believes that the news media professionals need tsunami training, because they have to report information to the public accurately and rapidly. People listen to and read news reports, often accepting them as fact. If a reporter doesn't report events clearly or offers misinformation, then people may be put in danger.

After the tsunami warning of October 1994, one television reporter said he couldn't understand why there was so much excitement about some potentially three-foot waves. "Three-foot waves," he said, "occur all the time in Hawai'i." As Walt points out, that reporter would have reacted quite differently if that three-foot wall of water were coming down his street.

Educating the public about tsunamis also means educating those who are responsible for public safety. Civil defense is charged with the responsibility of alerting the public and protecting them when a natural disaster is imminent. If a natural disaster such as a hurricane approaches Hawai'i, civil defense can get up-to-the-minute information about the hurricane's wind speed, direction, duration, size, and other important facts. That precise scientific data allows them to plan appropriate evacuation measures in a timely fashion. Most civil defense personnel are volunteers, trained to provide information, set up shelters and evacuation points, and coordinate many tasks related to a potential disaster. Although they are not scientists, they need to

Civil defense evacuation sign in Hilo, Hawai'i. Photo by Walter Dudley.

understand the details of every kind of natural disaster, including tsunamis.

Walt says, "We do periodic flights throughout the Hawaiian Islands with the civil air patrol. We take photographs all around the islands so that we can update our information about dangerous areas every few years. We also take photographs so that we can make maps of safe areas for people. In the aftermath of most tsunamis, scientists will go in with their tape measures and survey transits and figure out where the waves went. They'll also figure out the force and direction of the waves so that they can understand more about these powerful forces. That is valuable information, but I think it is far more important to keep people from being there when a tsunami happens."

Since tsunamis can't be tracked by radar or spotted by aircraft like hurricanes can be, the job of civil defense workers is often complex. They must sometimes issue alarms with minimal information. Frequently, they have to make educated guesses about the intensity of a tsunami. Despite all their good intentions, they are sometimes wrong. After several false alarms, the public may not believe or heed their warnings. This is why Walt spends a lot of time working with civil defense throughout Hawai'i.

Walt is aware of the sensationalization of tsunamis. Newspaper reporters, television announcers, and radio disk jockeys sometimes focus exclusively on the death and destruction of natural disasters. They often ignore the ways in which citizens could prevent some of that destruction or save their own lives. Walt says, "One of the things we fight with media people is the sensationalism. They want to know how many people died and how big the waves were. That's often all they care about. They don't want to know how a family can save themselves. I think that's much more important."

Walt believes that the best information about tsunamis is the information that comes *before* a tsunami strikes, not the facts and figures *after* a tsunami. Walt subscribes to the old saying, "An ounce of prevention is worth a pound of cure." In many ways, he is trying to prepare people as much as he is trying to inform them.

For Walt, communication and education are necessary and vital parts of his work. The public, those protecting the public, and those informing the public need to be aware of the potential dangers of tsunamis. "It's an interesting and frequently difficult challenge," Walt says. "But in 1960, people died because of misinformation broadcast over the radio, and I don't want to see that happen again."

TSUNAMI = TIDAL WAVE

U.S. DEPARTMENT OF COMMERCE
National Oceanic and Atmospheric Administration
National Weather Service

IF AN EARTHQUAKE OCCURS...

AND YOU MUST HOLD ONTO SOMETHING TO KEEP FROM FALLING...

HEAD FOR HIGH GROUND...

AS STRONG EARTHQUAKES CAN CAUSE TSUNAMIS

A tsunami warning poster informing people what to do when an earthquake occurs. Courtesy of NOAA.

8 The Pacific Tsunami Museum

"The local history of tsunamis in Hawai'i," says Walt, "contains many true stories of tragedy, sacrifice, and heroism, as well as accurate descriptions of tsunamis. It is the power of these true stories as told by the survivors themselves which has the ability to capture the imagination and educate audiences of residents and visitors who most need to understand the danger of tsunamis."

In 1988, Walt and his friend Min Lee wrote a book about tsunamis. In it, they included a note inviting readers who had experienced the terror of tsunami waves to contact Walt and tell him their stories. One woman, Jeanne Johnson, wrote Walt a letter saying she had a story to share about how she and her family had survived the tsunami in Hilo in 1946. She also said the stories of other tsunami survivors might be important and suggested the stories be recorded and preserved in a museum. At that time, there was no permanent collection of survival stories; thus began the idea for a tsunami museum.

Meetings were held and ideas were shared. An office for the museum was set up in a shopping center. Professional consultants were contacted, and a board of directors was elected. In an ironic twist of fate, the first meeting for the museum's board of directors was scheduled for October 4, 1994. On that day, though, an earthquake north of Japan generated a tsunami that raced across the ocean, and the first meeting to establish a museum dedicated to tsunami education was canceled following a statewide tsunami warning.

Later, after several successful meetings, the group, led by Walt and Jeanne, came to two conclusions. First, the museum's mission should be twofold: to preserve the oral history of Hawaiian tsunamis and to develop education programs to prevent loss of life. Second, the creation of a museum would take a lot of money: nearly four million dollars, by one estimate!

Just when the organizing group was about to lose hope of raising that kind of money, First Hawaiian Bank donated a building in downtown Hilo, in May 1997. The board of directors was thrilled. The building would become the homesite for the museum. But now the real work began.

The First Hawaiian Bank building in Hilo, home of the Pacific Tsunami Museum. Photo by Malika Dudley and Emily Dudley.

"There is a lot of talent in the community and people are so giving of their time, especially for this cause. People are really touched by what happened here and what these people have been through and the fact that it can happen again unless they do something. We did a lot of planning and a lot of recruiting, so that we could open a state-of-the-art museum and tell the whole story," says Walt.

Walt's dream culminated when the doors to the Pacific Tsunami Museum officially opened in June 1998. Since then, the community has continued working to assemble an impressive collection of exhibits and outreach programs. The museum, a tribute to those who lost their lives in past tsunamis, includes film and video presentations, computer simulations and virtual-reality pro-

One of the many displays inside the Pacific Tsunami Museum. Photo by Walter Dudley.

grams. There are a children's center, a video camera focused on Hilo Bay, and exhibits on the myths and legends of tsunamis, as well as a museum shop and several public-safety programs for visitors and residents alike.

One of the most touching exhibits is a collection of notebooks containing essays written by local students. The students had interviewed grandparents and elderly neighbors, inviting them to share their stories and memories of past tsunamis. Many of the survivors had never talked about those events before. Museum visitors frequently stop to read these memorable, often tragic recollections. One of the museum's volunteers said, "People will sit down to read these stories and they get so choked up."

Bunji Fujimoto, a survivor of the 1946 tsunami at Laupāhoehoe School, talks to a group of children at the Pacific Tsunami Museum. Photo by Walter Dudley.

One day an elderly man walked into the museum and began to look at the photographs and read the stories. He was so moved that he told Walt about his own experience during the tsunami of 1946. "There I was in the river. I had my mom on my back. A small boat came by, so I grabbed the painter [rope] to the boat. A kid washed by, so I threw him in the boat. I was really afraid. I had been in World War II and rescued some people. But this was different. This was hell," he said.

The museum assists with historical tours of Hilo and Laupāhoehoe for visiting groups of tourists. The museum also has several teachers' training programs in the summer, which draw educators from across Hawai'i to design and create tsunami-related lessons. Videos, photographs, and printed materials are provided for teachers to use in their classrooms.

The history of tsunamis in Hawai'i comprises many true stories of heroism, sacrifice, and tragedy. Walt continues to collect stories, adding them to the museum's archives, a "living" memorial to those who died in past tsunamis, and an education center for the prevention of casualties in the future.

The Laupāhoehoe Exhibit

This exhibit in the Pacific Tsunami Museum commemorates the teachers and students who lost their lives at Laupāhoehoe School during the tsunami of April 1, 1946. It is filled with photographs, stories, and other artifacts such as the following examples described and illustrated in the figures.

In one exhibit that includes paddles from the rescue effort, there is an oversized photograph of two people in a life raft; beneath the photo is this harrowing tale:

> *When a tsunami struck on the morning of April 1, 1946, Herbert Nishimoto was swept out to sea from Laupāhoehoe Point. The fifteen-year-old youngster was tossed around in the debris and wreckage for many hours. That afternoon, the Coast Guard dropped a rescue pack, which included a raft, two pad-*

dles, and some survival gear. As Herbert floated along the coast, he picked up two other boys. The coastline was hazardous and covered with steep cliffs. Herbert remembered camping at Pelekunu Bay around the other side of the North Kohala horn. He kept the raft close to shore and headed for the horn. When the boys reached Niue Bay a woman on shore spotted the raft and called the police. The police sent signal flares ablaze to alert the Coast Guard planes. Those on shore yelled for the boys to bring the raft into Niue Bay. They were fearful that if the boys passed the horn they would be swept out to sea. It was rough going, but the raft made it to shore. The boys had been at sea for two days. Herbert was injured, but all three boys survived. These are the paddles they used to bring the raft ashore at Niue Bay.

Laupāhoehoe Point as it looks today. Photo by author.

Waves crash off the end of Laupāhoehoe Point. Photo by author.

The following quote is printed on a wall of the exhibit:

> *The water pulled me under and when I came to the surface I found myself on the reef. Then I saw another wave coming. It sucked me under and dragged me out to sea. When I came up I couldn't see anyone around.*

Hanging on another wall is a beautiful handmade quilt, and the following sign is posted nearby:

> STUDENTS DONATE QUILT TO TSUNAMI MUSEUM OUT OF RESPECT FOR THOSE WHO LOST THEIR LIVES AT LAUPĀHOEHOE POINT DURING THE APRIL 1, 1946 TSUNAMI. *Students from Mrs. Sue Nozaki's home economics class at Laupāhoehoe Elementary and High*

School finished this special quilt. The quilt commemorates the twenty-four students and school faculty who lost their lives at Laupāhoehoe Point during the morning of April 1, 1946. The colors of the quilt represent the school's colors: blue and gold. On December 19, 1997, the quilt was donated to the Pacific Tsunami Museum.

The quilt that commemorates the teachers and students who lost their lives at Laupāhoehoe School on April 1, 1946. Photo by author.

One Day at the Museum

You can never fully understand the power of tsunamis until you look into the eyes of some of the survivors and realize the depths of their losses. I don't think anyone should ever have to go there again. That's what the museum is all about.

— *Walt Dudley*

One day, a woman came into the museum to look at the displays and exhibits. She stood for a long time in front of the Laupāhoehoe exhibit. A museum volunteer went over to talk with her, and the woman told this story.

On the morning of April 1, 1946, a young boy decided he did not want to go to school. Perhaps he'd had a premonition, but his mother insisted that he go. He left the house twice and came back both times to say that he didn't want to attend that day. Each time he came back, his mother told him that he must go to school.

Finally, the boy grabbed his jacket and walked down the road that led to Laupāhoehoe School. That was the last time his mother saw him alive. He was one of the students swept out to sea by the power of the tsunami. The family was devastated.

A few days later, the smell of wild orchids filled the house though it wasn't the season for orchids. The boy and his sister, who had been very close, had often gone together to pick wild orchids when the flowers were in bloom. The boy had always placed the orchids in his sister's hair, something he loved to do very much.

When the sister told the mother about this smell of orchids, the mother believed it was a sign from her son. Though she still grieved, the smell of orchids, the boy's favorite flowers, told her he was okay. The boy had died in the terrible tragedy at Laupāhoehoe, but the smell of his favorite flowers lingered in the house. His mother could be at peace.

The woman who told the story to the museum volunteer was the boy's sister. She had not told this to anyone until she saw the display in the museum more than fifty years later.

People's Lives Are Far More Important

> Studying tsunamis gives me a chance to . . . help people. . . . I would like every kid to know as much about tsunamis as they know about fires or hurricanes. I don't want them to be part of a town or village that is completely devastated. In the long run, it's people's lives that are far more important.
>
> — *Walt Dudley*

Unlike hurricanes or tornadoes, tsunamis aren't seasonal events. North Atlantic and Caribbean hurricanes, for example, form during the months of August, September, and October. Hurricanes also occur in the Pacific Ocean. Tornadoes are a major threat in the Midwest during the months of April, May, and June. But we've learned that tsunamis have no season and often no warning signs.

The Hawaiian Islands have experienced many natural disasters. Recent ones include Hurricane Iniki (ee-NEE-kee), which swept over the island of Kaua'i on September 11, 1992, devastating more than half of it. This hurricane swept in off the Pacific Ocean, packing winds of 130 miles per hour. There were gusts up to 160 miles per hour.

The eruption of Kīlauea (key-la-WAY-ah) volcano began in 1983 and has been continuously erupting ever since. Earthquakes and a swelling of the volcano preceded the initial eruption. In later years, a number of small villages and communities were buried under several feet of molten lava. There were no human deaths, because certain geological events were predicted and people were warned.

Unlike other natural disasters, Walt says, "A tsunami has the ability to cause tens of thousands of miles of destruction, along with tens of thousands of deaths—with absolutely no visible warning signs. That makes them unique, and it also makes them particularly fascinating. It's a beautiful day at the beach, the sun is out, and the ocean is calm. All of a sudden, the water withdraws, and you hear this horrible rumbling and experience an onslaught of water that just never stops."

For the people of Hawai'i, tsunamis come with the territory. For Walt, it's important the people understand the phenomena, to know they are deadly, and to comprehend their unpredictability, because no two tsunamis have ever been alike.

If the warning system in Hawai'i works properly, then people can be alerted ahead of time. This is particularly true for distantly

Boy Scout Troop 77 just before their hike to Halape on Thanksgiving weekend, 1975. Photo courtesy of the Dorothy H. Thompson Collection, Pacific Tsunami Museum.

generated tsunamis—those that may originate in Japan, Alaska, or South America. Unfortunately, the warning system isn't helpful for locally generated tsunamis, such as the one that struck one Thanksgiving weekend on the Big Island of Hawai'i.

On that weekend, thirty-four campers, including members of a Boy Scout troop, had decided to hike down to the beach at Halape (hah-LAH-pay), a remote area on the southern coast of the Big Island. Early on the morning of November 29, 1975, the campers were jolted awake by earth tremors. Some large rocks fell from a nearby cliff. But after talking about the quake for a while, the campers all went back to sleep. A little more than an hour later, a second and more powerful earthquake (7.2 on the Richter scale) struck the area. The epicenter of the quake was in the ocean, less than fifteen miles from Halape. The shaking of the earth again woke the campers. For several minutes, enormous boulders cascaded down the face of the mountain behind the campers. Then there was silence.

Suddenly, a five-foot-high wall of water crashed through a nearby stand of coconut palms. Campers sprinted from the beach and up the slope of the mountain. The rushing water tore shelters and tents apart. Some campers were entangled in bushes while others were thrown into a large ditch. The first wave slowly receded into the ocean. Shortly afterward, the second wave—an incredible twenty-five feet high—smashed into the makeshift community. Everything in its path was ripped apart and carried as far as three hundred feet inland. Trees, rocks, and people were tossed like they were lightweight toys. Nothing was spared. Later waves, much smaller than the first two, continued to wash over the landscape. At daybreak, the area looked as though it had been inside an enormous washing machine. Amazingly, only two people died, though nineteen suffered extensive injuries.

The two offshore earthquakes had been a signal. But several campers, frightened by the boulders rumbling down the mountainside after the first earthquake, had thought they would be safer moving their tents closer to the ocean. When the second, more powerful, earthquake struck, more campers scrambled toward the

The Halape palm grove after the earthquake and tsunami of 1975. Photo from the U.S. National Park Service, provided by the National Geophysical Data Center.

shoreline. Everyone wanted to escape the threat of falling rocks. Nobody realized that the true danger would come from the sea.

"Unfortunately, a warning system isn't very helpful for locally generated tsunamis," says Walt. "People need to know what to do. If the ground shakes, it could be an earthquake or it could be an earthquake that generated a tsunami. Either way, don't take any chances; it's time to move to higher ground. If the water withdraws, or it's acting funny, then that's also a sign to move to higher ground. People are often their own best warning systems."

Being respectful of tsunamis rather than scared of them comes from a knowledge about how they are generated. Walt's emphasis is not on how many buildings are destroyed or how big the waves are, but on how lives can be saved.

Tsunami Safety

Whether you live along a coastline or far inland, it's important to know about tsunamis. Sometime in your life you will probably travel to the ocean. You may visit a seashore, drive along a coastline, or take a trip in a boat. As Walt says, "It's important that all kids and adults know how the ocean behaves. Then they'll know what to do when the ocean is acting weird. We all know what to do when we see a thunderstorm approaching. We should also know what to do at the ocean. Knowing how nature works will help protect families, because they'll understand enough to take care of themselves and their loved ones."

Here are some safety tips and suggestions, whether you live close to the ocean or are just visiting.

Before a Tsunami: If you live near or are visiting an area where tsunamis could occur, there is a lot you can do in advance. Talk with members of your family and develop a "family tsunami plan." Make sure everyone knows how the warning system works and what it means. If your home is in a tsunami inundation area, select a safe place inland where your family can go. Use a map to trace at least two different routes to the "safe zone," in case of traffic congestion. If you live in or are visiting Hawai'i, check the front section of the telephone book for the tsunami evacuation maps. Practice tuning a portable radio to the emergency broadcast system so that you can obtain emergency or evacuation information. Be sure to have a small, easily portable first-aid kit available. Practice with the materials in the kit at least twice a year. Purchase a flashlight with good batteries. You may wish to contact your civil defense agency to obtain important evacuation and safety information for your area ahead of time.

During a Tsunami: Tsunamis are dangerous events and should not be taken lightly. A tsunami can be a series of waves that last for eight hours or more. Waves in a tsunami may be a hundred miles or more apart. The later waves are often larger than the earlier waves. Never go or return to a coastal area after the first two or

CIVIL DEFENSE
Tsunami Evacuation Maps
Evacuate all shaded areas.

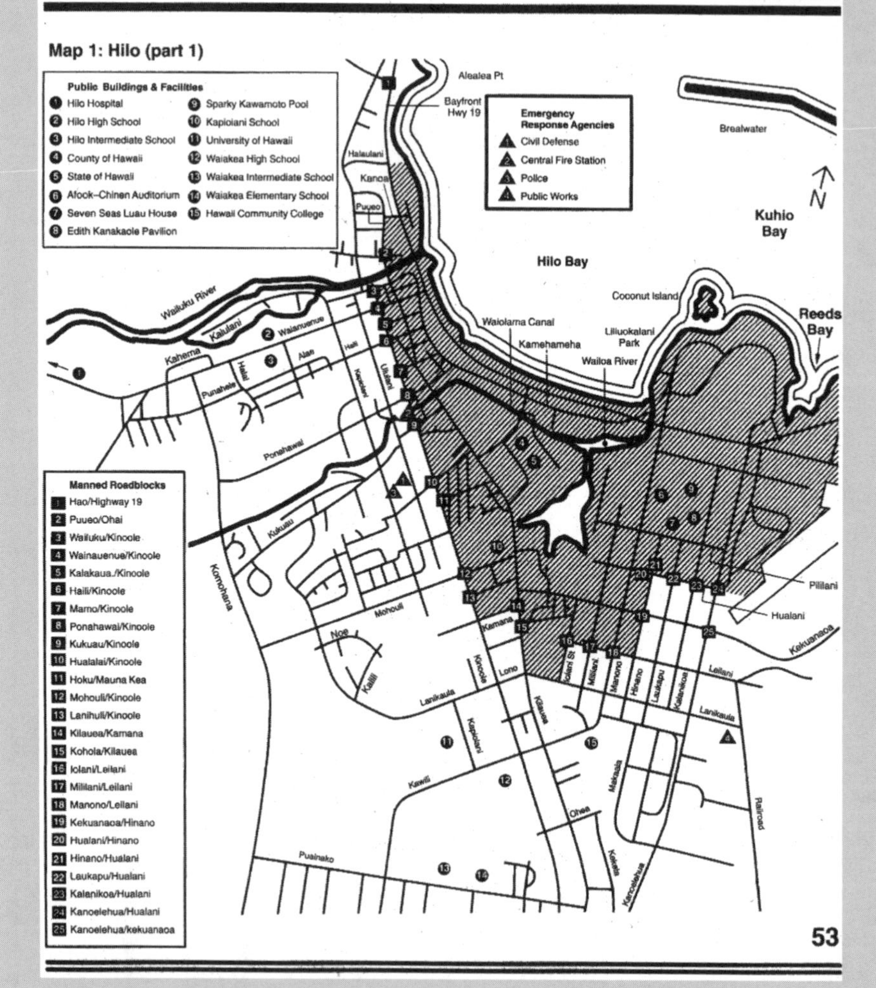

A civil defense tsunami-evacuation map that appears in the front pages of the Hilo, Hawai'i, phone book. Map by Verizon Hawaii.

three waves have hit. There may be more on the way! Stay away from any inundation areas and remain on high ground. Listen to radio broadcasts and return only when instructed to do so by the proper authorities.

After a Tsunami: When a tsunami has passed through an area, there may be hazardous conditions, such as downed power lines, broken gas lines, and floating sewage. Many buildings may have been damaged and be in danger of collapsing. Never go inside a building, or even near one, that has been hit by a tsunami. Never use the electricity or gas in your home until they have been checked by trained professionals. Call a plumber if you suspect that the plumbing may be damaged in any way. Check all food supplies carefully for water damage or contamination. If in doubt, discard anything that got wet. Be sure to have your drinking water checked by the local health department. Do not enter a tsunami zone unless instructed to do so by the authorities.

Downtown Hilo in the aftermath of the 1946 tsunami. Photo by the University of California at Berkeley, provided by the National Geophysical Data Center.

A True and Constant Threat 10

Hawai'i is struck by more tsunamis than any other region in the world, yet less than half the one million permanent residents and virtually none of the six million annual visitors to Hawai'i understand the tsunami hazard.

— *Walt Dudley*

Walt stands on the beach watching the panorama before him. Sunlight dances across the water, creating a magical iridescence. The wind is lightly blowing and the sea is gently rippled. Tall palms arc out over the bay as if bowing to its beauty. The scene is peaceful and serene, just another day in paradise.

As enchanting and majestic as life in Hawai'i is, it is not without its dangers. Created by volcanic eruptions over millions of years and constantly worn down by the erosion of wind and water, the islands have been shaped and influenced by meteorological, geological, and oceanographic forces for eons. These forces continue to influence life in Hawai'i.

Tsunamis are one of nature's most unpredictable phenomena, occurring every year in varying degrees somewhere in the Pacific Ocean. Since 1925 in Hawai'i, ten tsunamis have caused 223 deaths and more than one hundred million dollars in reported property damage. The April 1, 1946, tsunami was the worst natural disaster to strike the state of Hawai'i—even worse than Hurricane Iniki in 1992. The city of Hilo suffered the greatest

Walt Dudley on the beach. Photo by Michael Childers.

loss of life, reporting 96 of the 158 deaths, and over twenty-six million dollars (1946 figures) in property damage. The tsunami of 1960 resulted in the deaths of 61 people and property damage estimated at fifty million dollars.

Walt knows that another tsunami will strike the islands. His work and the work of other scientists around the world indicate that the danger is always there. What Walt and the others don't know is when that potentially destructive series of waves will strike. Nobody can predict an undersea earthquake or volcanic eruption. It could be next week or next year.

Part of the challenge with an unpredictable catastrophic

Part of the city of Hilo after the tsunami of 1946. Notice the enormous boulders swept inland by the power of the waves. Photo courtesy of the Aleta V. Smith Collection, Pacific Tsunami Museum.

event such as a tsunami is educating the public. Walt enjoys helping residents and tourists throughout the Hawaiian Islands appreciate and be prepared for these natural disasters, especially by collecting stories and sharing them.

"I really like being in contact with tsunami survivors," Walt says. "I feel so honored just to talk to these people. To be able to turn around and do something good with the stories is the greatest thrill. I like to know that I am doing the right thing."

Herbert Nishimoto, a survivor of the April 1, 1946, tsunami that struck Laupāhoehoe School, and Walt Dudley. Photo by Jeanne Johnston.

Whether listening to a tsunami survivor on the beach, hopping aboard a plane and interviewing a family in Honolulu, or reading diary entries of a former Hilo resident, Walt is always gathering stories. And if there is a way the stories can help save lives, Walt will include them in the exhibits at the Pacific Tsunami Museum or share them some other way with the people of Hawai'i.

You never know when a tsunami will strike. Walt wants you to be ready.

Things You Can Do

One of the best ways to learn about tsunamis is to visit your public library. Most librarians can assist you in locating books, illustrations, photographs, and other information about tsunamis from around the world. Some of the information may be technical or advanced, but the reference librarian, your teacher, or your parents can help you understand some of the complex materials.

"Surfing" the Internet is another way to obtain important information about tsunamis. Here are some Web sites you may want to check out:

1. http://www.germantown.k12.il.us/html/tsunami.html
   Created by a school in Illinois, this site has some up-to-the-minute information on tsunamis.

2. http://www.pmel.noaa.gov/tsunami
   This site has sophisticated information about mapping efforts, modeling and forecasting, frequently asked questions, and important tsunami events and data.

3. http://www.geophys.washington.edu/tsunami/welcome.html
   This site provides general information about tsunamis. Included is information on the impact of tsunamis, recent tsunami events, and the tsunami warning system.

4. http://library.thinkquest.org/J003007/Disasters2/menu/menu1.htm
   This site includes background information on tsunamis, the warning system, and a tsunami myth. You can also hear the story of a tsunami survivor at this site.

5. http://www.usc.edu/dept/tsunamis
   Here you can find maps of tsunamis around the world, links to other sites and video footage and animated simulations of tsunamis.

6. http://www.tsunami.org
   At this Web site for the Pacific Tsunami Museum, you can obtain the latest information and most current scientific data about tsunamis as well as the work the museum is doing to promote tsunami education throughout the Hawaiian Islands.

You may also want to contact the National Marine Educators Association (http://www.marine-ed.org) to get information about careers in marine science. Additionally, local colleges or universities with marine science programs will be able to provide you with important and valuable information on oceanographic careers.

There are two excellent videos available in most video stores or directly from the producers:

1. *Killer Wave* (catalog no. G51904) is produced by the National Geographic Society (1145 Seventeenth St., N.W., Washington, DC 20036; http://www.nationalgeographic.com). This video showcases the destructive power of tsunamis around the world. In addition, you can see and hear Walt Dudley describing these powerful waves.
2. *Raging Planet: Tidal Wave* (catalog no. 72333-80737-3RE1), is a video from the Discovery Channel (BMG Entertainment, 1540 Broadway, New York, NY 10036; http://www.discovery.com), which offers an inside look at how people around the world are learning to deal with these natural occurrences.

Glossary

Caldera A crater formed by an explosion or powerful volcanic eruption.

Crest The highest point of a wave.

Epicenter The place on the earth's surface that is directly above an earthquake's focus or starting point.

Inundation zone An area from the shore inland that is covered by water during a tsunami.

Magma Hot, molten rock under the earth's surface.

Richter scale A scale that measures the amount, or magnitude, of energy released by an earthquake.

Ring of Fire A circle of active volcanoes around the Pacific Ocean.

Run-up The vertical height above sea level that seawater flows onto land, which might be as much as seventy-five feet.

Subduction zone Giant ocean trenches where the tectonic plates that make up the earth's surface collide and are forced to overlap one another.

Surf The constant pounding of waves on the shoreline.

Tectonic plates Massive slabs of rock beneath the earth's land and sea.

Tidal Wave The rise and fall of ocean tides on a shoreline. They usually occur in twelve-hour cycles. The term is sometimes used erroneously as a name for a tsunami.

Tides The rise and fall of the sea due to the pull of the moon and the sun. During high tide, the water piles up along coastlines. During low tide, it moves away from coastlines.

Trough The lowest point of a wave.

Tsunami Long, fast, powerful waves caused by earthquakes, volcanoes, or landslides near or under the sea.

Tsunami warning An alert telling people that a tsunami is approaching and that they should move to a safe place.

Tsunami watch An alert letting people know that a tsunami may be forming and that they should listen for further news.

Wave height The distance from the trough of a wave up to its crest.

Wave length The distance from the crest of one wave to the crest of the next wave.

Wave period The time it takes two wave crests to pass the same point or marker. This measurement is used to estimate wave speed.

Wave train A series of waves going in the same direction.

Further Reading

Children's Literature

Fredericks, Anthony D. *Exploring the Oceans: Science Activities for Kids*. Golden, Colo.: Fulcrum Publishing, 1998.

Petty, Kate. *I Didn't Know That Tidal Waves Wash Away Cities*. Brookfield, Conn.: Copper Beech Books, 1999.

Souza, D. M. *Powerful Waves*. Minneapolis, Minn.: Carolrhoda Books, 1992.

Thompson, Luke. *Natural Disasters: Tsunamis*. New York: Grolier, 2000.

Adult Literature

Dudley, Walt. *Tsunamis in Hawaii*. Hilo, Hawai'i: Pacific Tsunami Museum, 1999.

Dudley, Walt, and Scott Stone. *The Tsunami of 1946 and 1960 and the Devastation of Hilo Town*. Virginia Beach, Va.: Donning Company Publishers, 2000.

Dudley, Walter, and Min Lee. *Tsunami!* 2d edition. Honolulu, Hawai'i: University of Hawai'i Press, 1998.

Index

About the Author

Anthony D. Fredericks is a former elementary classroom teacher now working as a professor of education at York College in York, Pennsylvania. He grew up in Southern California, where he spent his summers surfing, swimming, and sailing. He has written more than eighty books, including teachers' resource books and children's books. His children's books focus on animals, environmental concerns, and scientific topics for kids of all ages, including *Cannibal Animals, Under One Rock, Exploring the Rainforest,* and *Elephants for Kids.* Many of his books have won special awards and citations. Every year he visits many schools around the country as a visiting children's author. He is a frequent visitor to Hawai'i, and in his free time he enjoys snorkeling, hiking, and exploring. Readers can learn more about his school visits and books by logging on to www.afredericks.com.